SNOW

Robert Bolognesi

Understanding

Observing snow

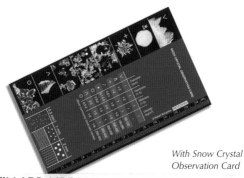

*With Snow Crystal
Observation Card*

Translation
Blyth Wright

To Frédérique

'Mirror, mirror on the wall,
Who is the fairest one of all?'
'The snow, my queen,
the driven snow.'

Thanks to Bruno Bouillard, Othmar Buser, Armand Dussex,
Fabrice Meyer and François Sivardière for their kind help.
Thanks also to Blyth Wright for his valuable contribution.

Introduction

Snow has no lack of admirers. As soon as winter arrives skiers, snowboarders, climbers and hill-walkers rush to the presence of the beautiful and enigmatic White Lady. But she can be a cruel mistress: on average each winter a hundred people succumb to her fearful rages, meeting death by avalanche.

For an avalanche to occur, the terrain must be suitable. However, this is obviously not the only requirement. The instability of the snowpack is also a crucial factor, and this depends very much on the physical make-up of the snow. Unfortunately this is quite difficult to determine – not least because in the mountains the snowpack varies with both place and time. Assessing the state of the snow is therefore always challenging, but it is, nevertheless, an essential skill for anyone who wants to travel the mountains in winter and enjoy them safely.

What is snow? How is it formed? How does it change? Where and when is it dangerous? This book attempts to provide some answers to these questions. It describes the process of formation and evolution of snow, and shows practical methods of examining and analysing snow cover. It enables readers to identify the many forms which snow may take, and to assess avalanche risk more precisely and reliably.

A. Snow crystals form in the atmosphere or on the ground and take many different forms.

B. On the ground, snow crystals undergo changes due to mechanical and thermodynamic processes.

C. The changes affecting the snow modify its physical properties, particularly its cohesion.

D. The snowpack is made up of successive snow layers of varying cohesion: it is therefore stratified. The stratigraphy of the snowpack affects its stability.

E. The weakness of the snowpack may be estimated initially from continuous meteorological observation.

p. 34 and 35

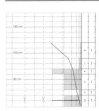

F. The weakness of the snowpack may be shown by means of snow profiles.

p. 36 to 47

G. The weakness of the snowpack may be observed by means of stability tests.

p. 48 to 54

The validity of measurements and tests is localised. The snowpack situation is only one element in assessing avalanche risk.

p. 55 to 58

Understanding snow

HOW SNOW IS FORMED

Water is always present in the atmosphere in the form of water vapour, an invisible and odourless gas. But the colder the air is, the less water vapour it can hold. A fall in atmospheric temperature (which occurs as air rises, for example) may therefore lead to the **condensation** of water vapour and so to the formation of clouds.

The process of the formation of snow begins in the cloud with the birth of a minute ice particles: this is called **nucleation**. In pure water, nucleation (referred to as 'homogenous' nucleation) does not take place until the water reaches about -40°C; so water can remain liquid at temperatures below freezing in a state known as '**supersaturation**'. But in a cloud, in the presence of freezing nuclei (mineral or organic particles from 0.1 to 10 **microns**), nucleation (known as 'heterogenous' nucleation) usually takes place at higher temperatures (below -10°C, but it can still occur up to about -3°C).

Starting with a germ of ice, a snow crystal develops by **sublimation** and **deposition**, sometimes with the added freezing of surrounding water droplets. This development takes place through the growth of the faces, sides or points of the germ, and leads to the formation of crystals which are prism-, plate- or star-shaped. The snow crystals continue their growth at the expense of smaller crystals until they reach a mass at which they are too heavy to remain suspended in the cloud. When they have obtained this 'critical' mass (which is greater if the cloud contains strong turbulence), they fall, more or less joined into snowflakes. If the atmospheric temperature remains sufficiently cold all the way down to ground, then we have a snowfall (the snow–rain limit is normally about 300m below the level of the **0°C isotherm**.)

Note: technical terms in bold are explained in greater detail in the Glossary on pp. 60–61.

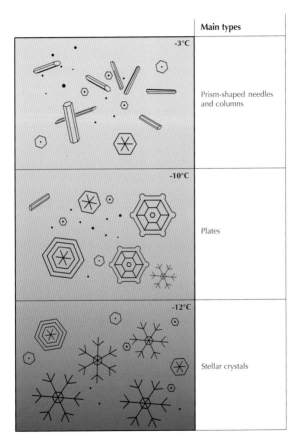

	Main types
-3°C	Prism-shaped needles and columns
-10°C	Plates
-12°C	Stellar crystals

Formation of snow in the atmosphere (basic forms).

Types of snow crystal.

The variety of snow crystals

The form and size of snow crystals varies considerably depending on the atmospheric conditions (temperature, humidity and turbulence) in which they are created. Several hundred distinct forms, derived from the basic types, have been catalogued. The World Meteorological Organisation (WMO) has divided snow crystals into ten large families. In order to predict likely changes in newly fallen snow, it is important to take into account the snow's initial crystallography, distinguishing especially between **dendritic** crystals, which generally have a branched form, and round crystals, which are much larger.

Types of snow crystal		
Form	**Name**	
	Stellar (star-shaped) crystals	
	Plates	
	Columns	
	Needles	
	Spatial dendrites	
	Capped columns	
	Irregular crystals	Dendritic crystals
	Graupel	
	Ice pellet (sleet)	Rounded crystals
	Hail	

Classification of solid precipitation according to the WMO system.

Example of a dendritic crystal: star-shaped (symbol: +).

Dendritic crystals

The commonest dendritic crystals are star-shaped crystals, which form when the temperature within a cloud is low (below about -12°C). It is easy to see them with the naked eye during snowfalls in calm weather, when they fall gently to the ground individually or assembled into light snowflakes. But there are many other less branched forms of dentritic crystals, such as needles and columns as well as all kinds of irregular particles. On the ground, snow made up of these kinds of crystals is subject to major changes.

Example of a rounded crystal (graupel) (symbol: ⚥).

Rounded crystals

These are mainly seen in the form of **graupel**, which is common in the mountains in winter. Graupel crystals are easily recognised by their quite rounded form and fairly large size: their diameter is often more than 5mm. They are formed in turbulent clouds with large vertical development (typically large cumulonimbus), and are created by water droplets freezing onto the crystals before they fall. This obscures their original form and gives them their characteristic appearance – like small polystyrene beads. They often fall during brief but violent thunder showers. They do not change much once within the snowpack, except by melting.

Surface hoar crystals (symbol: ∨).

Surface Hoar

If the layer of air in contact with the snowpack is moist, and if the surface of the snowpack is subject to strong cooling, this leads to the formation of surface hoar crystals. These may become very large (several centimetres) and often resemble fern leaves in shape. They can often form quite thick layers when skies are completely clear. So this type of snow does not fall from the sky, but forms on the ground! As with rounded crystals, surface hoar crystals undergo little change when they are in the snowpack and will remain there until melting takes place.

Atmospheric riming.

Atmospheric riming

Atmospheric riming (generally known as 'fog crystals') appears when there is thick cloud along with windy and cold weather. It takes the form of deposits of compact white ice, made up of air and frozen droplets. In dense supersaturated cloud these deposits build up into the wind on crags and other objects. Atmospheric riming, which is common in regions of wet–cold climate and in tropical high mountain areas, also occurs in the Alps, where it can form a sheath of ice on rock faces. It can also cause serious damage to infrastructure, especially to aerials and electricity cables, which may collapse under the load.

HOW SNOW CHANGES

In the course of their journey through the atmosphere and during their stay on the ground, snow crystals are continuously subject to mechanical (physical) influences and energy (thermal) fluxes that change their structure.

Mechanical influences

At high altitude, snow is frequently carried by the wind, both during and after snowfall. Even a wind as light as 15km/hr can move snow lying on the ground. This movement occurs through the rolling, sliding and **saltation** of the snow crystals, or sometimes through the turbulent suspension of crystals ('spindrift') in the air when winds are strong. As they are moved, the snow crystals undergo air friction as well as multiple collisions. Thus, they shatter into tiny particles not exceeding a few tenths of a millimetre in size. These form dense deposits on the ground that rapidly gain in **cohesion**. There are three types of deposit created by the wind – **cornices**, **cushions** and slabs – and these are often quite compact. Light winds accompanying snowfall can also form snow deposits, which are soft and of various densities.

Within the snowpack, crystals are also subject to strong influences due to loading. A crystal of new snow, for instance, lying 10cm below the surface, bears a hundred times its own weight. The fine-branched dendritic crystals rapidly break up under this load. In this way **settling** of the snow occurs even as it is deposited.

These natural mechanical influences particularly affect deposits of dendritic snow that have not yet been subject to much change. Their effect is to reduce the size of the crystals and to pack them together, thus increasing the density and the cohesion of the snow layer.

Wind transport of snow with turbulent suspension.

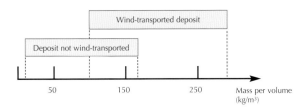

Effect of wind on the mass per volume of fresh snow.

Energy flux

The snowpack constantly gains and loses energy.

Gains are due to:
- direct and diffused solar radiation
- infrared radiation emitted and reflected by clouds and the atmosphere
- changes in phase of water (condensation)
- energy fluxes from the soil (flux of released energy and **geothermal flux**).

Losses are due to:
- infrared radiation from the snow
- changes in phase of water (evaporation, sublimation).

Rain and snowfall can cause the snowpack to gain or lose energy, depending on their respective temperatures.

Finally, the difference in temperature between the snow surface and the air in contact with it also causes gains or losses of energy which may be amplified by wind.

The various gains and losses of energy at the snow surface can be of the order of several hundred watts per square metre (w/m^2). The geothermal flux is much weaker (only a few w/m^2).

The energy balance of the snowpack represents the difference between the gains and losses of heat during a given period of time. It may be positive or negative, and results in a raising or lowering of the mean temperature of the snowpack and large variations in the temperature of the snow surface, which usually fluctuates between 0°C and -30°C.

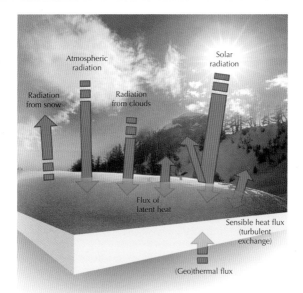

Energy flux at the surface and the base of the snowpack.

The variations in energy balance and the thermal inertia of the snowpack lead to the creation within the snow layers of a vertical **temperature gradient**. This is defined as the difference between temperatures measured at two points on the same vertical in relation to the distance which separates them: for example, a difference in temperature of 5°C for two points 20cm apart would give a temperature gradient of 25°C/m. Thermodynamic changes in snow vary according to the value of this thermal gradient. Like mechanical influences, they lead to a densification of the snow, but do not always contribute to an increase in its **cohesion**.

The metamorphism of snow: general points

There are three types of **metamorphism**, two of them relating to dry snow and one to wet snow.

Dry snow:

- *Dry snow metamorphism* (formerly known as 'equitemperature metamorphism' or 'destructive metamorphism') occurs when the vertical temperature gradient within the snowpack is below about 5°C/m. This may happen when there is a long period of mild, cloudy weather, particularly after prolonged snowfall. In these circumstances, water vapour flows in all directions from the convexities and is deposited as ice in the concavities of the snow crystals;

- *kinetic growth* (formerly known as 'temperature gradient metamorphism' or 'constructive metamorphism') takes place when the vertical temperature gradient is between 5°C/m and 20°C/m (moderate gradient), or when it exceeds 20°C/m (strong gradient). Cold, clear and calm weather leads to this type of metamorphism, particularly in shaded areas such as north-facing slopes. Transfer of material from grain to grain, by sublimation and deposition, then takes place upwards through the snowpack.

Wet snow:

- *melt metamorphism* takes place if water penetrates the snowpack. It occurs, therefore, after wet weather or warm and/or sunny periods.

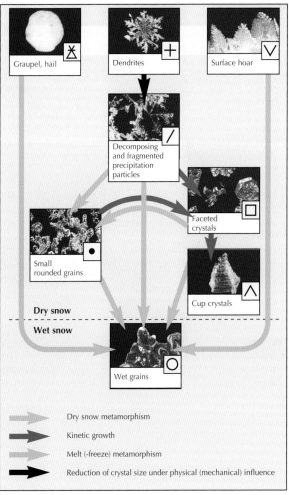

The metamorphism of snow.

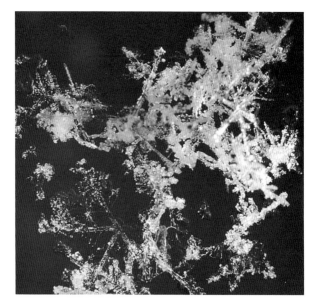

Decomposing and fragmented precipitation particles (symbol: /).

Dry snow metamorphism

This type of metamorphism occurs in those layers of dry snow made up of recognisable particles or even faceted crystals. It comes about through the **sublimation** and **deposition** of water vapour from the convexities of crystals to their cavities. This transfer of material causes crystals to become uniform and reduces them in size, while angular forms disappear. The small radii of curvature of these crystals encourages the formation of ice bridges at their points of contact.

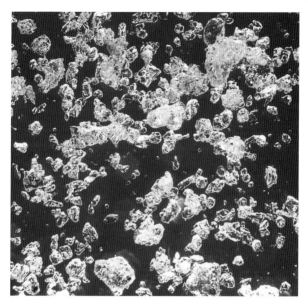

Rounded grains (symbol: •).

The end result of dry snow metamorphism under weak temperature gradient is small rounded grains. These make up 'packed'
snow, which is often weight-bearing and shows good cohesion.
This is the snow of pisted ski slopes, created by mechanical
packing. This type of snow is ideal for constructing igloos,
because it is dry and is easily cut into blocks which bind together
well. However, this snow also forms the slabs that easily create
fractures, and thus has the potential to cause huge avalanches.

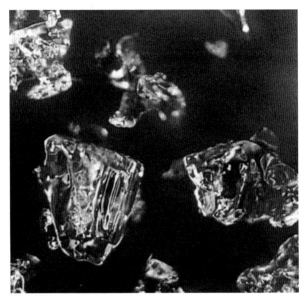

Faceted crystals (symbol: □).

Kinetic growth

When there is a large difference in temperature (a vertical temperature gradient greater than 5°C/m) between the bottom and top of a layer of snow that has remained dry since falling, a very particular type of metamorphism occurs in the snow crystals. In effect, they rebuild themselves, taking on an appearance very different from that of the star-shaped crystals of new snow: they become angular with flat faces, then like little pyramids of striated ice, sometimes several millimetres in size. These very characteristic grains are known as 'cup crystals' or 'depth hoar'.

Cup crystals, also known as 'depth hoar' (symbol: ∧).

Cup crystals form snow layers which, although quite dense (200–400kg/m³), have poor cohesion and create fragile zones within the snowpack. These crystals form when the weather is cold, dry, cloudless and calm, particularly at the start of the winter, when solar radiation is weak, the snowpack is still thin and the nights are long. At high altitude, layers of cup crystals are often found at the base of the snowpack on north-facing slopes. Cup crystals may form after a few days and will not disappear until affected by thaw.

Wet grains (symbol: O).

Melt metamorphism

The presence of water in the snowpack, following rain, heavy thaw or prolonged sunshine, initiates a major change within the snow crystals, which grow in size and become wet grains. These are quite large (0.5mm–2mm) and are always clustered, joined to each other by a film of water or matrix of ice. This kind of snow is dense (400kg/m³ on average), looking rather dull because of its low **albedo**, and is well known to skiers as 'spring snow'. This metamorphosed snow shows very low resistance (strength) when saturated, but very high resistance when refrozen ('melt–freeze cycle'). This is the final stage of snow's evolution before it disappears through thaw.

Cohesion due to re-freezing: when the snow looks like ice…

During a period of fine weather, the energy balance of the surface layer of the snowpack goes through considerable variation. The contrast is particularly marked in springtime, when the snow becomes very wet during the day and refreezes at night. This melt–freeze cycle leads to the formation of thicker and thicker crusts of large rounded grains. The metamorphism of the snow is at this point greatly influenced by topography: the state of the snow surface varies widely according to slope aspect. On the other hand, when the snow becomes wet due to rising temperatures or a period of rain, layers of wet grains appear on slopes of all orientations.

Mass per volume	50-500 kg/m³
Diameter of grains	0.1-5 mm
Latent heat of fusion	334 J/g
Latent heat of sublimation	2 834 J/g
Calorific capacity	2.09 J/g/°C
Thermal conductivity	0.05-1 W/m/°C
Albedo (visible radiation)	0.5-0.9
Albedo (infrared radiation)	approx. 0
Resistance to compression	4-400 kPa
Resistance to traction	2-200 kPa
Resistance to shearing	0.1-300 kPa

Some current values typical of snow (taken from a variety of sources).

Properties of snow

The properties of snow, which greatly depend on its state of metamorphism, are very variable: its mass per volume may vary by a ratio of 1:10; the diameter of grains by a ratio of 1:50; and its resistance to shearing by a ratio of 1:1000. Variations of this order may sometimes be seen in the course of a single day or, at a given moment, within the same massif: if each snow property had its own colour, the winter mountains would present a multi-coloured spectacle!

Amongst the snow properties that determine stability, **cohesion** is undoubtedly the easiest to monitor.

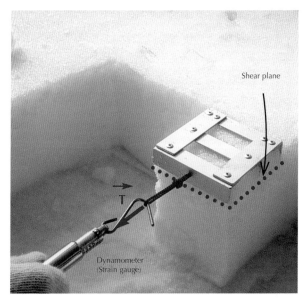

Measuring resistance to shearing using a shear frame, which allows the cohesion of the snow to be ascertained.

Cohesion of snow
The cohesion of snow is the result of the forces linking the grains within it. It depends on the number and the nature of these links. Thus, it is possible to identify different types of cohesion because they result in certain ranges of values for the mechanical resistance of the snow: **settling** assembles new snow crystals; **sintering** unites small rounded crystals; **capillarity** or refreezing links large rounded grains.

In the field, it is possible, if not to quantify the cohesion of the snow precisely, to measure or estimate its resistance to shearing, and so to identify weak layers within the snowpack.

Cohesion by settling, typical of dry new snow.

Cohesion by settling

Settling results from the entwining of the branches of star-shaped crystals. It occurs in fresh snow that falls in calm conditions, that has not changed much, is not packed and that remains dry. Resistance values for this type of cohesion are extremely low.

Settling gives the snow layer a precarious and ephemeral equilibrium, which, when it fails, leads to trees losing their snow load and typically causes single-point loose surface releases on steep slopes, during or just after the snowfall.

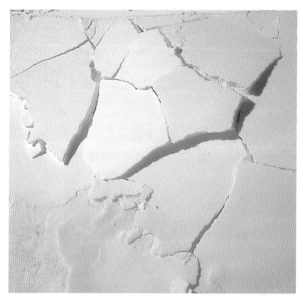

Cohesion by sintering leads to slab formation.

Cohesion by sintering

Sintering takes place when ice bridges form between snow crystals, particularly when crystal size is small. It occurs in dry snow that has been subject to dry snow metamorphism. This type of cohesion produces bonded snow and fairly hard slabs. Under these circumstances natural avalanches are not very common, though this is not the case for triggered avalanches. This kind of snow favours the development of fractures due to the load caused by skiers, which can be the cause of very large avalanches.

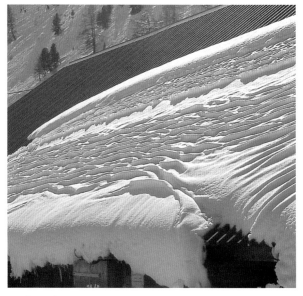

Capillary cohesion sometimes gives the snow a paste-like appearance.

Capillary cohesion

Capillary cohesion is associated with wet snow, in which the snow grains are held together by the film of water that surrounds them. This kind of cohesion, while not very strong, is enough to prevent natural avalanche release as long as the liquid water content is not too great. However, cohesion is considerably diminished in very wet snow, especially when **percolation** has begun. This is when many avalanches release, first of all on steep, sun-exposed slopes.

Cohesion by refreezing.

Cohesion by refreezing

When liquid water contained in the snowpack freezes under the effect of a cooling of the snowpack, the grains become held in a matrix of ice which acts like cement and gives the snow very strong cohesion: cohesion by refreezing. This process forms crusts, which may vary in thickness from a few millimetres to several centimetres. In the latter case, the snow cover becomes very strong. If it involves the surface snow layer, the probability of avalanche is virtually nil.

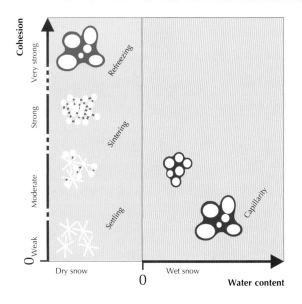

Cohesion of different types of snow.

Cohesion, crystallography and snow stability

The strength of a snow layer depends on the kind of cohesion that holds its grains together. But although the number of natural avalanches decreases as the snow gains in strength, the probability of triggered avalanches may conversely be increased, as there will be a period when cohesion is still too weak to prevent fractures but strong enough to create them, thus giving rise to slab avalanches. Snow of moderate cohesion can thus sometimes be more dangerous, being more unpredictable, than snow of very poor cohesion.

Form	Cohesion
•	moderate
V	very poor
O	very good
□	poor
O	moderate
□	poor
O	moderate
∧	very poor
O	moderate

Section of a snowpack which has undergone a succession of contrasting meteorological situations.

Stratification of the snowpack

Each of the snow layers deposited on the ground is subject to a particular meteorological history that determines its evolution and its characteristics. The snowpack is the sum of these different layers: it thus shows quite a visible stratification. Its make-up evolves throughout the winter, and alters more quickly and radically when local weather conditions are extreme and changeable. The snowpack can differ substantially from one place to another: for example, the snowpack can be twice as deep on one part of a mountainside as it is on another.

Observing snow

In the mountains, the state of the snow cover can be interpreted by assessment, measurement or tests. Assessments help to identify local snowpack variations; measurements assist in establishing conditions on particular aspects and at certain altitudes; and tests can pick out local instabilities.

ASSESSMENTS

Continuous assessment of weather conditions and an understanding of how the snowpack evolves often make it possible to estimate to some extent the condition of the snowpack in a given place. So, it is relatively easy to diagnose the formation of unstable accumulations and the destabilisation of surface layers, which are the two main causes of snowpack failure, as well as the presence of weak layers and the development of sliding surfaces, which are the principal factors that tend to increase instability.

Formation of accumulations

Heavy snowfalls and wind transport of snow always lead to the formation of unstable accumulations. This instability can appear, on slopes of more than 30°, as soon as the depth of the new snow exceeds 30cm. The instability increases as the intensity of the snowfall increases. It is worth noting that, in places, snow transported by the wind can form, in only a few hours, accumulations comparable to those resulting from a snowfall of several days!

Destabilisation of surface layers

This destabilisation often takes place after rain, prolonged thaw or strong sunshine. It can be very rapid if the snowpack is dense and wet: it only needs a few hours of rain or spring sunshine to destabilise the snow cover.

Presence of weak layers

Weak layers, made up of faceted grains and cup crystals, are formed by the metamorphism of dry snow crystals in situations where the radiation balance of the snowpack stays strongly negative. So, be alert to the possible presence of these weak layers on all shaded slopes whenever the weather stays clear, dry, cold and calm. Weak layers may also be made up of surface hoar or graupel. Each of these is easy to observe....

Presence of sliding surfaces

These surfaces are essentially melt–freeze crusts on which succeeding layers of snow may slide. Sliding surfaces form when strong cooling affects a snowpack in which the surface layer has been moistened by rain or solar warming. As with weak layers, they do not become a concern until buried by new snow layers.

A little notebook can prove very useful for recording daily weather observations, which, even in summary, can provide a very good resource for analysing snow conditions.

MEASUREMENTS

Ram penetrometer profile

The use of the ram penetrometer features among the classic techniques of snow observation. It seeks to quantify the resistance to penetration of the different layers in the snowpack, by measuring the penetration depth of a vertical tube driven in by a sliding hammer.

The tube is in three graduated sections, which may be joined, and above this is a guide, also graduated, on which the hammer slides. Based on the penetration (d) of the tube, the number (n) of the drops of the hammer (P) from the height (h) chosen by the operator, and the mass (qQ) of the tube, the resistance (R) of the snow layer may be determined by a simple formula:

$$R = (P.n.h\ /d) + q.Q + P$$

Resistance measurements are supplemented by a temperature profile as well as a description of the crystallography, the hardness, and sometimes also the shear resistance and liquid water content of each layer in the snowpack.

All these observations are then recorded in a graphic, providing a full picture of the internal composition of the snowpack at the point of sampling.

Ram profiles are usually carried out by trained observers on behalf of avalanche forecasting services. On a mountain route, it is difficult to perform a profile like this, as it requires quite heavy and bulky equipment (although it can be carried in a rucksack). However, it is useful to know how to interpret the information, as it can be found in the offices of most ski patrols, who generally take the observations once a week.

Doing a ram profile.

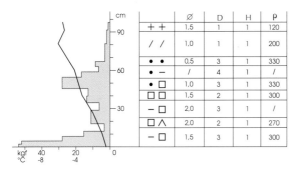

		Ø	D	H	ρ
+	+	1.5	1	1	120
/	/	1.0	1	1	200
•	•	0.5	3	1	330
•	—	/	4	1	/
•	□	1.0	3	1	330
□	□	1.5	2	1	300
—	□	2.0	3	1	/
□	∧	2.0	2	1	270
—	□	1.5	3	1	300

Example of a drawn-up ram profile, with stratigraphy and temperature profile
(D = hardness; H = humidity).

Simplified profile: principle

The test profile enables the internal structure of the snowpack to be shown by a histogram of the resistance of the snow to tangential shearing (a very significant indicator of the weakness of the snowpack).

The strength of the different snow layers is estimated according to their crystallography and their hardness, which is evaluated by a simple hand-hardness test. This is based on certain statistical correlations which are known to exist for each type of snow (particularly for snow of poor cohesion) between the hardness and the penetrative resistance of each layer.

So this method does not require a penetrometer, only a shovel, a pencil, a knife and a recording form to illustrate the results.

A thermometer can also be added to this simple snow kit to provide temperature information.

The advantage of this system is that it gives relevant information (enabling thin weak layers to be detected) with great rapidity; so it is possible to perform several observations at different places in the same area, which gives a much more reliable idea of the actual snow conditions. In addition, it does not require any special equipment, so it can be carried out in the course of a climb. The profile can be drawn up on the spot, thus immediately giving a clear picture of the snowpack structure. The profile can also be communicated to other people, for instance by displaying it at the nearest mountain hut.

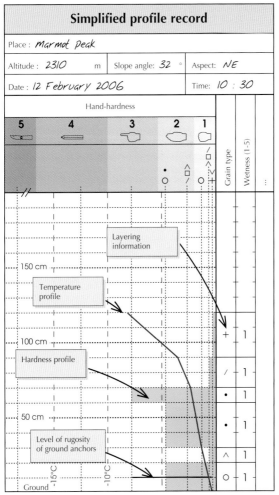

Test profile record

Test profile: method

For a test profile, dig a pit to ground level in a place not threatened by avalanche. This place should be chosen so as to be representative of the suspect slope. Identify the different snow layers according to visual indications on the back wall of the pit. Then, for each layer:

1. Assess the hardness by hand-hardness test.
   Remove the snow lying on the deep layers of the snowpack for the hardness test to be valid (otherwise hardness may be over-estimated).

Penetration		Hardness	Value
Fist		Very soft	1
Four-finger maximum		Soft	2
One-finger maximum		Quite hard	3
Pencil maximum		Hard	4
Knife maximum		Very hard	5

Hand-harness test of a snow layer.

2. Assess the wetness by snowball test.

Observation	Wetness of snow	Value
No snowball	Dry	1
Dry glove	Moist	2
Wet glove	Wet	3
Water drops	Very wet	4
Flooded	Slush	5

Manual test for wetness of a snow layer.

3. Identify the crystallography by scraping some grains from the back wall of the pit, putting them on a knife blade, looking at them through a magnifier and taking into account all the observable criteria: grain type and diameter, hardness and wetness of the layer.

4. As necessary, take the snow temperatures at the bottom and top of the layer.

Taking the snow temperature.

5. Enter the shear resistances on the graph (worked out from the crystallography and hardnesses – formula available from METEORISK), as well as the temperatures.

If the profile is to be filed or published, don't forget to put in the time and location data.

The example below shows the recording of the penetrative resistance of a thick layer of wet grains of hardness 1, with the base of the layer at 1.4m from the ground (this is the surface layer of the snowpack).

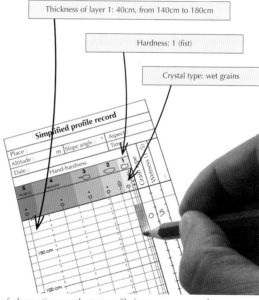

Thickness of layer 1: 40cm, from 140cm to 180cm

Hardness: 1 (fist)

Crystal type: wet grains

Record of observations on the test profile form.

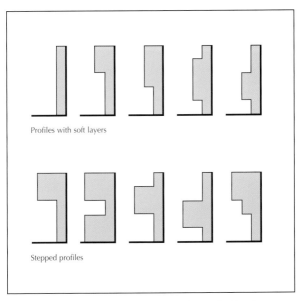

Profiles with soft layers

Stepped profiles

Examples of hardness profiles particularly suited to avalanche release.

Hardness profiles and stability

The stability of the snowpack may often be assessed from the observed hardness profile (as long as this is representative of the snowpack in the suspect area).

Some kinds of profile are frequently associated with avalanche release. This is especially the case with snowpacks which include:

- a thick surface layer of soft snow (profiles with soft layers)
- a weak layer of any thickness overlaid by soft or moderately hard snow (stepped profiles).

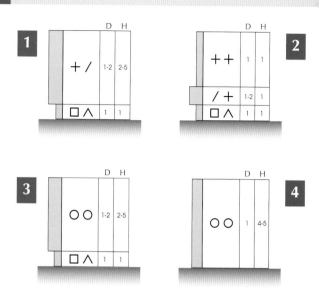

Examples of particularly unstable profiles with soft layers.

Profiles with soft layers

Soft layers are seen in two types of snowpack: fresh snow, not wind-blown and little metamorphosed (see profiles 1 and 2 above); and saturated snowpacks made up of wet grains in the process of melting (see 3 and 4). These snowpacks become more unstable with either the increasing thickness of the soft layer (and especially if it rests on a hard layer – or the ground, see photo opposite – providing a sliding surface, see profile 4) or the increasing cohesion of the snow where it overlies a weak layer (cup crystals, buried surface hoar or graupel, see 1, 2 and 3). The latter case is the cause of numerous off-piste accidents, the avalanche usually being triggered by the victims.

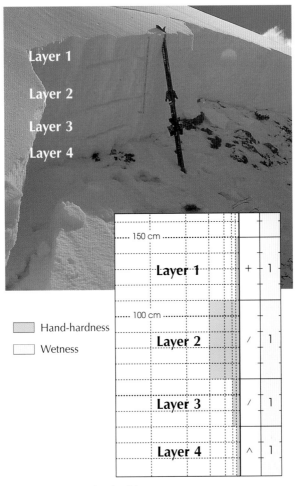

Fracture of a snowpack with soft layers.
Do not take observations on avalanche sites. Danger!

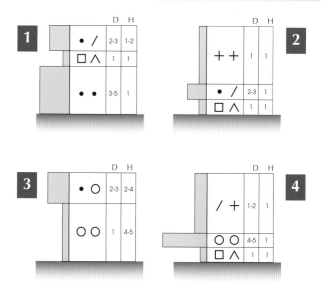

Examples of particularly unstable stepped profiles.

Stepped profiles

Stepped profiles are typical of snowpacks which show melt–freeze crusts or slab in their upper layers. With melt–freeze crusts (see profiles 3 and 4 above), the snowpacks are generally stable, even completely stable if the crust is thick and results from a prolonged thaw–freeze cycle. With slab (see 1 and 2), stability is much harder to assess; fractured slabs of varying degrees of hardness and thickness are seen. The size of the slab, the nature of the underlying layer, the topography of the gully (angle and shape), as well as the location and size of the anchors all become significant influences on stability.

Layer 1

Layer 2

Layer 3

150 cm

100 cm

Hand-hardness

Wetness

1 · ○ · 2

2 · 1

3 · ∧ · 1

Fracture of a snowpack with stepped stratigraphy.
Do not take observations on avalanche sites. Danger!

STABILITY TESTS

Tests of snowpack stability involve observing its resistance to various forces. They give an immediate measure of the localised stability of the snowpack and so provide a valuable complement to any global analysis, which they may confirm or, on the other hand, call into question.

Many stability tests exist. The most useful, beginning with the simplest, are the stick test, the shovel test and the sliding block (or wedge) test.

The stick test attempts to find weaknesses within the snowpack on the basis of how easy or difficult it is to push a ski stick into it. This test is fairly crude, but still permits weak underlying layers to be found, if they are not too thin, as well as sliding surfaces. So, it is not without interest, particularly as it is very easy to perform.

The shovel test aims to show possible shear planes within the layers of the snowpack, by isolating a column and pulling on successive layers with a shovel. This test is subjective and quite hard to interpret.

The sliding block test (also called 'Rutschblock') originated in Switzerland where it has been used for several decades, particularly by the avalanche service of the Swiss army and the Federal Institute for Snow and Avalanche Research at Davos, which uses it in the compilation of its daily avalanche bulletins. In performing this test, a block of snow is isolated from the snowpack and progressively loaded in order to promote failure. The instability of the snowpack is calculated from the magnitude of the load required to provoke the shear.

The sliding block test has two big advantages:
• the procedure is simple, precise and above all reproducible
• it requires only two very commonly available items: a shovel and a length of cord.

With the addition of an avalanche probe, it is possible to carry out, a little faster, a similar test: the sliding wedge (see p. 52).

General points when carrying out stability tests:
• choose a spot away from avalanche hazard! The test site should also be free from stone or serac fall. This point needs to be emphasised, as several accidents have demonstrated;
• choose a place which has as far as possible the same topographic features (slope angle, anchorages) and geographical nature (aspect, altitude) as the suspect slope. The test should therefore be performed on a slope of at least 30°;
• choose a site which is representative of the suspect slope in terms of snowpack structure;
• make sure, with the use of an avalanche probe, that the chosen spot does not have any large terrain anomalies (holes, rocks, streams, etc);
• adhere strictly to the test procedure, so that results from other people at different places may be compared;
• verify the result of the test, if necessary, by carrying out a second test in a nearby place, especially if the snowpack is uneven.

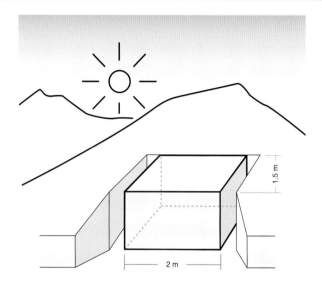

Preparation of the sliding block.

The sliding block test

1. Isolate an oblong block from the snowpack, digging to ground
   level on three sides and cutting the uphill side with a cord.

   Hint: when knotted every 50cm, the cord becomes a useful
   measuring and cutting tool.

2. Load the block progressively, looking carefully for any sign of
   failure: to this end, approach the block, skis on, then perform
   downsinks and jumps. If the block does not fail, finish with
   jumps without skis.

The sliding block test: loading.

The result of the sliding block test is expressed on a scale which takes into account the load necessary to promote failure.

Failure criteria	Value	Interpretation
During isolation	1	☠
Stepping onto block	2	**Danger**
On (rapid) downsink	3	
Jump	4	⚠
Second or third jump	5	**Caution**
Jump without skis	6	**The slopes**
No failure possible	7	**indicated are basically safe.**

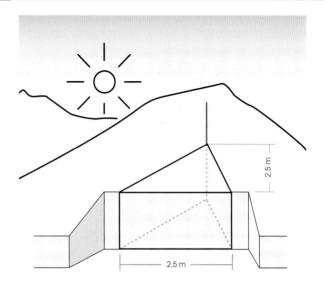

Preparation of the sliding wedge.

The sliding wedge test

1. Isolate a block from the snowpack, in the form of a triangular-based prism, digging to the ground on the front face and cutting the uphill sides using the cord running over the avalanche probe (see above).

2. Apply loading to the block (same method as for the sliding block), looking carefully for any sign of failure.

INTERPRETATION

General procedure for analysis

The field observations and tests provide, in local terms, valuable information on the snowpack. Unfortunately, bringing this together often requires time that is not always available during a climb. However, it is often enough to refer to the observations and tests only to confirm your assumptions about the likely effect on the snowpack of past and present local weather conditions.

So, the most efficient procedure for analysis is undoubtedly the following:

1. knowing the past state of the snowpack and the recent (local) weather conditions, formulate hypotheses regarding the snow layering likely to be encountered on different aspects and at different altitudes in the area concerned;

2. if possible, elaborate these hypotheses using the information in the avalanche bulletin;

3. confirm the validity of the hypotheses by appropriate field observations and tests. If the hypotheses differ greatly from observations on the ground, it would be advisable to carry out fuller observations and to revisit the analysis of the local avalanche hazard. On the other hand, if the observations confirm the hypotheses, there is no need to carry out further observations: it will be possible to base the risk analysis on the hypotheses.

Interpretation of snow profiles

Stacked and stepped profiles with low mean penetrative resistance are often associated with high instability, which is often widespread (with weak layers of cup crystals, wet grains or graupel generally existing on all aspects and at all altitudes similar to those of the test site). On the other hand, pyramidal profiles and those with high average penetrative resistance are typical of stable snowpacks. These remarks are not universally true, but profiles are still useful, at least for understanding the snowpack state and foreseeing its evolution.

Interpreting sliding block (or wedge) results

With occasional exceptions, the following rules, although very simple, seem helpful.

Value	Interpretation
1,2	The slopes indicated are very unstable (avoid!)
3	The slopes indicated are unstable (avoid except in emergency: retreat, rescue, etc)
4,5	The slopes indicated are suspect (negotiate carefully, with wide spacing)
6,7	The slopes indicated are basically safe

Stability tests have a spatial range which is even more restricted than that of snow profiles. So, the results of any test can apply only to slopes of the same altitude and aspect, and of the same angle, as the test site. Scrupulous consideration of the validity range for any stability test is therefore an essential part of its interpretation.

Limitations in interpreting data

Snow observations are susceptible only to an approximate interpretation because:

- the representativeness of data is limited, as much in space as
in time. This reality is awkward...but unavoidable. It means
that additional observations must be carried out if the snowpack has been disturbed (especially after a period of strong
winds) or is changing rapidly, because if observations then are
inadequate, they can lead to wrong conclusions;

- the precision of the data is very relative. Observations taken
without instruments are approximate and not always significant. So stability tests do not in general bring out instabilities
in very soft or very thick slabs, and are not suitable for very
wet or very incohesive snow;

- the number of observations is small. Describing a snowpack
by the use of several dozen variables is still a summary.

Nevertheless, despite their inadequacy, snow observations
enable an evaluation of snowpack stability to be perceptibly
refined. So they are worth making and using; but always bear in
mind that the more varied and more numerous your information
sources, the more reliable your analysis of a situation will be,
which emphasises the importance of observations being rapidly
made and shared.

In the mountains, it is often preferable to take into account a
blending of many approximate observations rather than a single
more precise measurement.

Snowpack information and avalanche risk

Measurements and tests on the snowpack enable us to understand snow conditions better and, particularly, to reveal the presence of internal weak layers, sliding surfaces or accumulations of poorly bonded snow.

However, it is important to remember that avalanche risk is not determined by snow conditions alone. It is also dependent on many other factors that contribute to the two components of avalanche risk: the probability of avalanche and the potential damage.

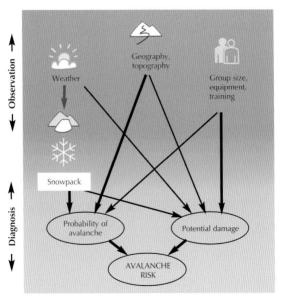

The snowpack state is only one of the elements taken into account when analysing avalanche risk.

Evaluation of avalanche risk
A brief reminder…

- The *probability of avalanche* is estimated from snow data (make-up of the snowpack), from topographical data (angle, profile, terrain roughness) and from information relating to the group (size, training, equipment) as its effects on the snowpack may trigger the avalanche. Note that knowledge of recent weather patterns (precipitation, wind and wind transport, temperature and cloud) can be a substitute for snow observations in the first instance.

- The *potential damage* of an avalanche is also dependent on snowpack factors (volume and type of snow available), weather conditions at the time (visibility), topography (terrain traps) and, again, the group (equipment, rescue training). In areas that are inhabited or where other people are present, the vulnerability of life, limb and property to avalanche must be taken into consideration.

The assessment of avalanche risk is therefore a difficult process of synthesis requiring a huge body of information, not limited to snowpack observation alone. A more detailed treatment of this subject is found in *Avalanche!* (Cicerone Press, 2007).

Test profile record

Place :

Altitude :   m | Slope angle :   ° | Aspect :

Date : | Time :

Hand-hardness					Grain type	Wetness (1-5)	:
5	**4**	**3**	**2**	**1**			

150 cm

100 cm

50 cm

−15°C   −10°C   −5°C

Ground

© Robert Bolognesi *METEORISK*

Useful websites

For further information...

American Avalanche Association
www.americanavalancheassociation.org

Australian Ski Patrol Association
www.skipatrol.org.au

Canadian Avalanche Association
www.avalanche.ca

Cyberspace Snow and Avalanche Center
www.csac.org

METEORISK
www.meteorisk.com

sportscotland Avalanche Information Service
www.sais.gov.uk

US Forest Service National Avalanche Center
www.avalanche.org

New Zealand Mountain Safety Council Inc.
www.avalanche.net.nz

Glossary

Albedo: fraction of solar radiation reflected by the snow.

Calorific capacity: energy required to raise the temperature of a body by 1°C.

Capillarity: a function of cohesion, adhesion and surface tension within the snow grain / water mix. As this depends upon molecular attraction within the water content, it is a potentially unstable state.

Cohesion: the forces uniting grains of snow.

Condensation: passage from the gaseous to the liquid state.

Cornice: accumulation of snow on a ridge or plateau rim, often overhanging, formed by wind.

Cushion: accumulation of drifted snow in a zone of wind deceleration.

Dendrite: branched development, star-shaped crystals.

Deposition: passage from the gaseous to the solid state.

Endothermic: absorbing heat.

Exothermic: emitting heat.

Geothermal flux: transmission upwards to ground surface of heat from warmer layers nearer the core of the Earth. Relatively insignificant compared to incident solar heating.

Graupel: large (around 5mm diameter), rimed, rounded crystals, resembling small polystyrene pellets, often seen during showers and thunderstorms.

0°C isotherm: theoretical plane formed by points where the atmospheric temperature is 0°C.

Isothermal: lacking significant temperature variablility.

Latent heat: energy absorbed or returned during a change of phase.

Metamorphism: changes in snow, arising from energy flux and mechanical influences, affecting its physical properties.

Micron: one thousandth of a millimetre.

Nucleation: the process of the formation in the atmosphere of an initial tiny ice crystal from the freezing of a water droplet around a nucleus which may be a tiny dust or salt particle. This is known as 'heterogenous nucleation'. In the absence of freezing nuclei, all water droplets freeze spontaneously at -40°C (homogenous nucleation).

Percolation: flow of water (from rain or thaw) towards the bottom of a snow layer, occurring when the liquid water content of the layer exceeds its capacity for retention.

Saltation: displacement, by successive jumps, of snow grains on the ground due to wind.

Settling: the linking of snow crystals by entanglement of their branches.

Sintering: the linking of snow grains by the formation of ice bridges at their points of contact.

Stratification: deposition in successive layers.

Sublimation: passage from solid to gaseous state.

Supersaturation: metastable state of a body which stays liquid at a temperature below its freezing point.

Temperature gradient: a measure of the difference in temperatures taken at two points relative to their distance apart, expressed in °C/m.

Index

© Robert Bolognesi, 2007

ISBN-13: 978 1 85284 474 5

Photography and Illustrations
© Robert Bolognesi

Translation
Blyth Wright

Robert Bolognesi, PhD, Polytechnic of Lausanne, is a ski patrolman, graduate of the Institute of Alpine Geography, Grenoble, former researcher at the Federal Snow and Avalanche Research Centre, Davos, avalanche control practitioner and founder of METEORISK, specialising in the amelioration of weather-related risks.

By the same author
Avalanche!
With *NivoTest*, avalanche risk calculation tool

Blyth Wright has climbed and skied extensively in the UK, the Alps and North America for over 40 years.

He was formerly Assistant Director of the International School of Mountaineering in Leysin, Switzerland, then worked for 20 years as an Instructor at Glenmore Lodge National Sports Centre. He has since 1989 acted as Co-ordinator of what is now the **sport**scotland Avalanche Information Service. He has represented Scotland and the UK at many international meetings of snow and avalanche workers. He jointly authored the popular book *A Chance in a Million? – Scottish Avalanches*.

Breaking a lifetime's habit, he visited the Himalaya in 2004 and climbed Ama Dablam. He hopes to continue with this interest.

2 POLICE SQUARE, MILNTHORPE, CUMBRIA LA7 7PY
www.cicerone.co.uk